WHOLE WIDI
GOLDEN GATE
BRIDGE

by Kristine Spanier, MLIS

pogo

Ideas for Parents and Teachers

Pogo Books let children practice reading informational text while introducing them to nonfiction features such as headings, labels, sidebars, maps, and diagrams, as well as a table of contents, glossary, and index.

Carefully leveled text with a strong photo match offers early fluent readers the support they need to succeed.

Before Reading

• "Walk" through the book and point out the various nonfiction features. Ask the student what purpose each feature serves.

• Look at the glossary together. Read and discuss the words.

Read the Book

• Have the child read the book independently.

• Invite him or her to list questions that arise from reading.

After Reading

• Discuss the child's questions. Talk about how he or she might find answers to those questions.

• Prompt the child to think more. Ask: Bridges help travelers get places faster. Do you have bridges near you? Do you know how long they are?

Pogo Books are published by Jump!
5357 Penn Avenue South
Minneapolis, MN 55419
www.jumplibrary.com

Library of Congress Cataloging-in-Publication Data

Names: Spanier, Kristine, author.
Title: Golden gate bridge / by Kristine Spanier, MLIS.
Description: Minneapolis: Jump!, Inc., 2022.
Series: Whole wide world
Includes index. | Audience: Ages 7-10
Identifiers: LCCN 2021027924 (print)
LCCN 2021027925 (ebook)
ISBN 9781636903101 (hardcover)
ISBN 9781636903118 (paperback)
ISBN 9781636903125 (ebook)
Subjects: LCSH: Golden Gate Bridge
(San Francisco, Calif.)–Juvenile literature.
Classification: LCC TG25.S225 S64 2022 (print)
LCC TG25.S225 (ebook) | DDC 624.2/30979461–dc23
LC record available at https://lccn.loc.gov/2021027924
LC ebook record available at https://lccn.loc.gov/2021027925

Editor: Jenna Gleisner
Designer: Molly Ballanger

Photo Credits: Luciano Mortula - LGM/Shutterstock, cover; Marti Bug Catcher/Shutterstock, 1; trekandshoot/Shutterstock, 3; Library of Congress, 4; Nick Starichenko/Shutterstock, 5; EpicStockMedia/Shutterstock, 6-7; gob_cu/Shutterstock, 8; Underwood Archives/Getty, 9, 10; Ernst Haas/Getty, 10-11; Radoslaw Lecyk/Shutterstock, 12-13; Sueddeutsche Zeitung Photo/Alamy, 14-15; tata_illustrator/Shutterstock, 15; Galen Rowell/Mountain Light/Alamy, 16; Cheryl Rinzler/Alamy, 17; CelsoDiniz/iStock, 18-19; Fast Speed Imagery/Shutterstock, 20-21; Allen Penton/Shutterstock, 23.

Printed in the United States of America at Corporate Graphics in North Mankato, Minnesota.

TABLE OF CONTENTS

CHAPTER 1

CROSSING THE GOLDEN GATE STRAIT

The Golden Gate **Strait** is in California. It lies between San Francisco and Marin County. In the early 1900s, people **ferried** across it. But ferries were slow. A bridge would make the trip much faster.

Golden Gate Strait

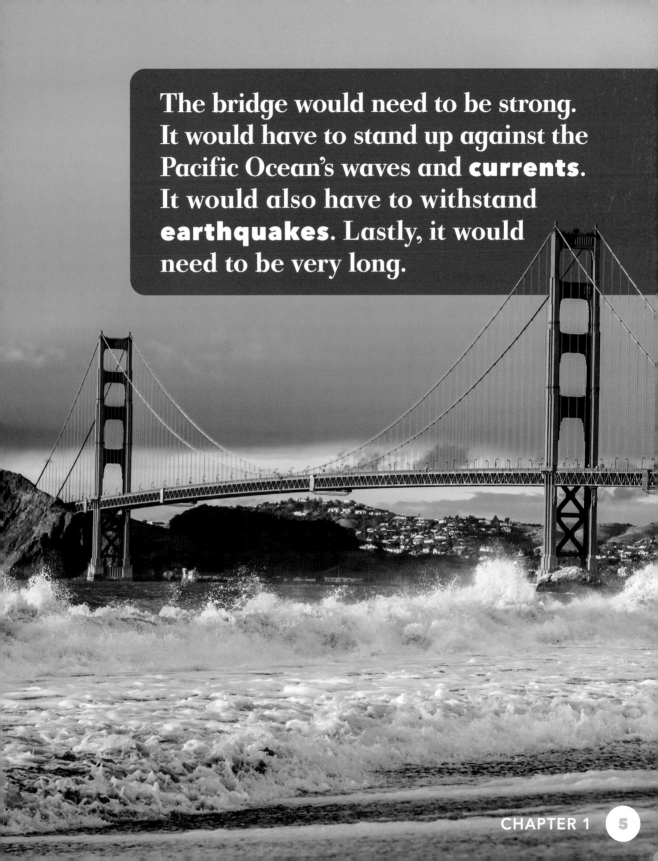

The bridge would need to be strong. It would have to stand up against the Pacific Ocean's waves and **currents**. It would also have to withstand **earthquakes**. Lastly, it would need to be very long.

Joseph Strauss and Charles Ellis were the lead **engineers**. They knew a **suspension bridge** could span the distance.

Construction on the Golden Gate Bridge began on January 5, 1933. It would be 8,981 feet (2,737 meters) long. Its towers would be 746 feet (227 m) tall. This would make it the longest and tallest bridge of its kind.

DID YOU KNOW?

The Golden Gate Bridge was the longest suspension bridge in the world for 27 years. As of 2022, the longest was the Akashi Kaikyo Bridge. It is in Japan. It is 12,831 feet (3,911 m) long!

tower

CHAPTER 2

BUILDING THE BRIDGE

The towers are made of steel. So are the two main cables. They hold up the roadway. They are very strong. But they are light enough to bend in the wind.

cable ·····▶

north pier

south pier

Two **piers** support the towers. The north pier was built close to shore. The south pier had to be set in deep water. Workers blasted a hole in the ocean floor. They poured concrete inside to hold the pier.

Building the bridge was dangerous. High winds could blow workers off. They buckled onto safety lines. They also wore hard hats. A safety net was placed below them. It saved 19 workers who fell off the bridge!

safety net

hard hat · · · · ▶

Irving Morrow was an **architect** on the project. He designed the towers. He chose the color orange. Why? This color stands out against the water and sky. It even stands out in thick fog. This way, boat captains could see it in any weather.

WHAT DO YOU THINK?

The U.S. Navy wanted to paint the bridge yellow and black. They thought these colors would be easier to see in fog. What color would you choose? Why?

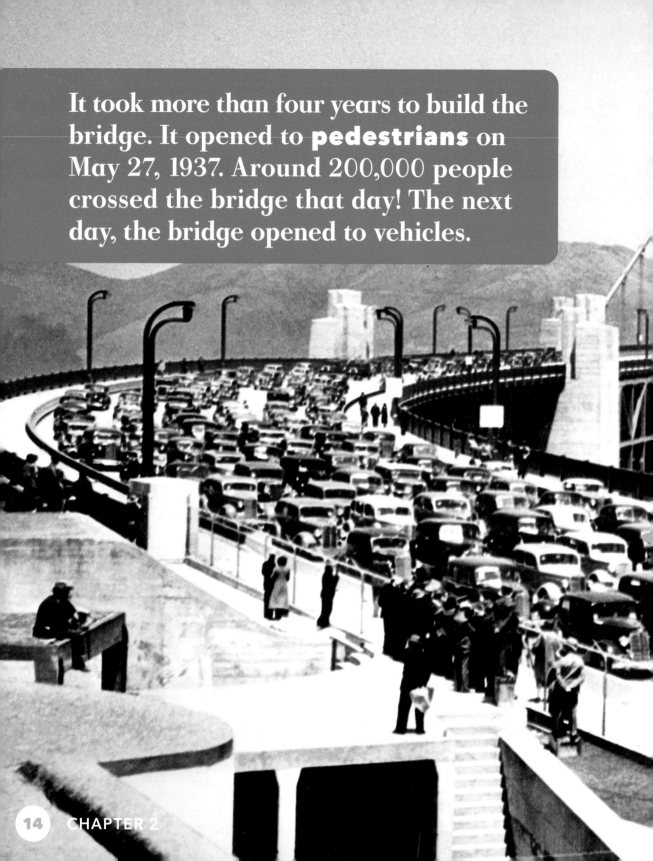

It took more than four years to build the bridge. It opened to **pedestrians** on May 27, 1937. Around 200,000 people crossed the bridge that day! The next day, the bridge opened to vehicles.

TAKE A LOOK!

What are the parts of the Golden Gate Bridge? Take a look!

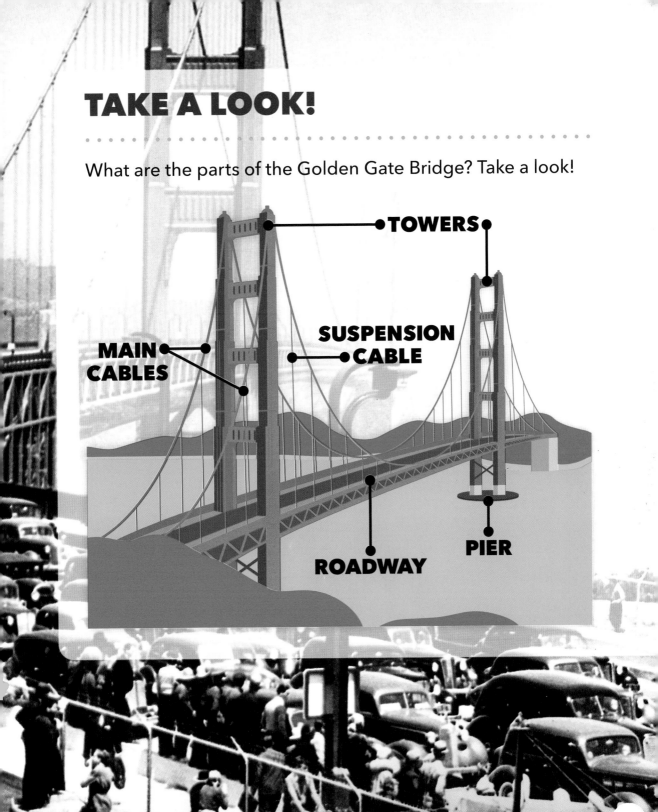

TOWERS

SUSPENSION CABLE

MAIN CABLES

ROADWAY

PIER

CHAPTER 3

THE GOLDEN GATE TODAY

Today, workers check the bridge daily. Why? They want to make sure it is safe. Salty air from the ocean creates **rust**. Workers paint the bridge every year.

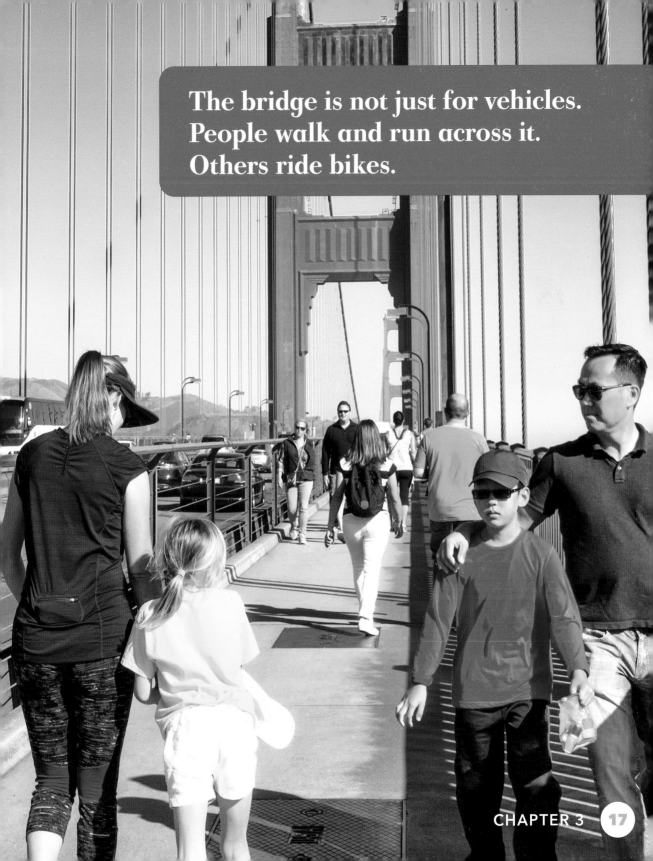

The bridge is not just for vehicles. People walk and run across it. Others ride bikes.

More than 100,000 vehicles cross the bridge each day. That is a lot of traffic!

WHAT DO YOU THINK?

Drivers must pay a **toll** to cross to San Francisco. This money pays to fix the bridge. Do you think this is fair? Why or why not?

$

The Golden Gate Bridge is a **symbol** of San Francisco. It is one of the world's most famous bridges. Have you ever crossed it? Would you like to?

QUICK FACTS & TOOLS

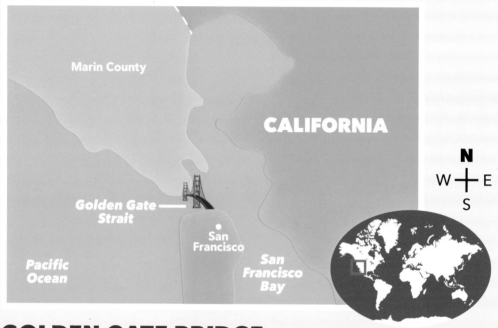

GOLDEN GATE BRIDGE

Location:
San Francisco, California

Length: 8,981 feet (2,737 meters)

Years Built: 1933 to 1937

Primary Engineers:
Joseph Strauss and Charles Ellis

Primary Architect:
Irving Morrow

Purpose: to connect
San Francisco and Marin County

**Number of Vehicle Crossings
Each Year:** around 40 million

GLOSSARY

architect: A person who designs buildings and supervises the way they are built.

currents: Movements of water in one direction.

earthquakes: Sudden, violent shakings of Earth that may damage buildings and cause injuries.

engineers: People who are specially trained to design and build large structures.

ferried: Traveled across a body of water on a ferry boat.

pedestrians: People who travel on foot.

piers: Pillars that support a bridge.

rust: A flaky, reddish-brown coating that forms on iron and steel when it is exposed to moist air.

strait: A narrow strip of water that connects two larger bodies of water.

suspension bridge: A bridge that hangs from cables or chains that are strung from towers.

symbol: An object that stands for, suggests, or represents something else.

toll: A charge or tax paid for using a highway, bridge, or tunnel.

INDEX

TO LEARN MORE

Finding more information is as easy as 1, 2, 3.

❶ Go to www.factsurfer.com

❷ Enter "GoldenGateBridge" into the search box.

❸ Choose your book to see a list of websites.

FACT SURFER